Driving Through Futures Past

Mid-20th Century Automotive Design

Hampton C. Wayt

Foreward by
Leslie M. Kendall
Curator
Petersen Automotive Museum

Driving Through Futures Past

Published by Kythe Publishing Co., Aiken, SC.

First Edition. First Printing February 2006.

Although the authors and publisher have made every effort to ensure the accuracy and completeness of the information contained in this book, we assume no responsibility for errors, inaccuracies, omissions or any inconsistency herein.

Photography by Lista's Studio of Photography, Aiken, SC, unless noted otherwise.

ISBN: 0-9755043-1-2
LCCN: 2006900004

This book is published as a record of the exhibit of the same title which was featured at the Petersen Automotive Museum from April 16th - September 11th, 2005.

Acknowledgements:

I would like to thank Leslie M. Kendall and Dick Messer of the Petersen Automotive Museum for making the Driving Through Futures Past™ exhibit possible, and those individuals and institutions who loaned material for this exhibit, including the Henry Ford Museum, Dave DiMartino, Gary MacNorius, Jimmy Valentine of the Metropolitan Pit Stop, Roger Willbanks, Terry Graboski, Daimler Chrysler, Larry Smith, Charles S. Hayes, and the Norman Bel Geddes Collection at the University of Texas at Austin. I would like to thank my friends Fred Sharf, Leo Brereton, Norman Brosterman, and Gary MacNorius for their support of this project and keeping me focused. And finally, I am forever indebted the dozens of designers and their families who have welcomed me into their homes over the years.

Finally, I would like to give a special thanks to my brother, Howard J. Wayt, Jr., for his many hours of dedication to the production of this catalog.

KYTHE PUBLISHING

www.kythepublishing.com

Design for a Streamline Motor Car by Ingemar K. Rystedt, illustrated by C.C. Bratten, 1925 (Exhibit #1)

Phantom Corsair Patent Model by Rust Heinz, modeled by Maurice Schwartz, 1936 (Exhibit #5)

Driving Through Futures Past: Mid-Century Automotive Concepts

Some of the most ingenious automotive designs of all time were conceived between 1925 and 1972. Imagining a future in which technology would provide a solution to every problem, mid-century designers conveyed their concepts of advanced automobiles though paintings, drawings, scale models and full-size prototypes. Regarded in today's cynical world as amusing, but unrealistic, these two- and three-dimensional works were conceived at a time of rapid scientific advancement that many believed would progress indefinitely. When the realities of limited nonrenewable fuel supplies, pollution concerns, and safety issues were acknowledged and modern engineers turned their attentions from the theoretical to the practical, the renderings, models, and cars were put away and all but forgotten.

In the intervening decades, technology has made many advanced ideas viable and it is fortunate that a large number of the remarkable renderings, models, and full-size prototypes have survived to inspire a new generation of artists and engineers. The Petersen Automotive Museum exhibition Driving Through Futures Past offers a glimpse into the creative processes that enabled designers a half-century ago to conceive such bold visions of our automotive future. From fanciful renderings of gyroscopic rocket cars and atomic-powered flying vehicles to hand-built scale models and full-size automobiles such as the ultra-streamlined 1936 Stout Scarab and 1954 Pontiac Bonneville dream car, these concepts embody technologies that remind us not only of how far we have come, but also how far we have yet to go.

Leslie M. Kendall, Curator
Petersen Automotive Museum

Image courtesy of the Earl Family Archive

Harley J. Earl (1893-1969)

This book is dedicated to the memory of Harley J. Earl, one of the most hotly collected artists of the 20th Century, whose name is ironically often unknown among collectors of his work. And also to the talented artist designers who followed in Earl's footsteps, many of whom are included in this catalog.

I: The Earl School

In his lifetime, General Motors President Alfred P. Sloan witnessed automobile companies evolve from glorified blacksmith shops into powerful corporations. He saw how precision manufacturing methods and standardization of parts made automobiles of all makes similar in quality and reliability. He also noticed how similar they were in appearance, sharing the same basic layout of four wheels, a motor in front, a radiator positioned between two headlights, and a passenger compartment with the driver's seat on the left. Looking to the future, Sloan realized that the most important thing that would distinguish the automobiles of his company from those of its competitors would be their aesthetics.

In 1927 Sloan sought out and hired Harley J. Earl, the son of a well-known California coachbuilder who had gained fame designing flashy vehicles for Hollywood stars. Sloan asked Earl to move to Detroit in order to assemble and direct a group of designers who would be responsible for creating a distinct look for each of the corporation's automotive lines. This new arm of General Motors was named the Art and Colour Section (later Styling Section) and was the first of its kind in the industry. With the creation of this studio, the profession of the automobile designer was born.

The philosophy of the automobile designer evolved directly from the on-the-job experiences of Harley Earl himself. One of the first observations Earl made upon arriving at GM was that engineers appeared to be more interested in what could not be done than in what could be done. Because of his frustration with this approach, Earl did not include the engineering department in styling decisions, believing that a good engineer could build virtually anything if properly motivated. Earl found that one of the best ways to motivate engineers was to fire any of them who claimed his designs were impossible to manufacture. The power Earl wielded meant that his designers needed no more than a passing understanding of engineering. Those who followed Earl's self-evolved philosophy of placing appearance considerations first before engineering concerns eventually became known as "automobile stylists" regardless of the field in which they worked. Industrial designer Richard Arbib, for example, was still referred to as an automobile stylist well into the 1990s even though he had not directly worked for an automobile company since 1942.

It would be more than a decade after Earl was hired by GM before schools to train automobile designers began to appear. As a result, in the early years of the Art and Colour Section General Motors recruited talent from wherever it could be found. Some early GM designers were engineers who had a keen interest in making automobiles beautiful as well as functional. A large number of recruits had been trained as professional artists for other fields. And some new hires arrived with degrees in industrial design, a field still in its infancy. No matter the background (and in dire contrast to today) the only fundamental qualifications for new employees were artistic talent and a passion for automobiles.

Many Earl-trained designers ultimately left GM and moved to other corporations. Earl's philosophy of design went with these designers and spread his influence. The training schools that were eventually established also followed the standards set by Harley J. Earl. And many of Earl's students (such as Gene Bordinat, Virgil M. Exner Sr., and Elwood Engle) came to be the driving forces behind the design studios of General Motor's competitors. They all had Harley Earl to thank for their great success.

By the 1970s, a mere fine arts degree no longer guaranteed professional success in the world of automotive styling. Because of a myriad of government regulations and modern ecological, economic, safety, and political realities, modern automobile designers must have engineering training. But while Earl's approach to design is now considered antiquated, the fundamental purpose of the automobile designer still remains as Earl envisioned it.

II: The Evolution of Automotive Style 1927-1970

The first noteworthy trend in the automobile styling profession, which came during the height of the Great

LaSalle Proposal by Paul A. Meyer, 1937 (Exhibit #7)

Depression of the 1930s, was streamlining. Inspired by a school of design that embraced essentials and rejected frivolity, streamlining came to represent the application of both science and art. The "science" of streamlining involved the application of aerodynamic principles to the exterior design of a car to reduce wind resistance, while the "art" of streamlining involved the smoothing of body panels and integration of exterior components in order to conceal machinery as attractively as possible.

The earliest application of scientific aerodynamic principles to motorcar body design came from aviation designers, whose work resulted in shapes that enhanced both speed and efficiency. In contrast, the artistic approach resulted in more unified designs, which often imparted an impression of speed. But since the primary purpose of the "art" of streamlining was not scientific in nature, actual performance improvements were merely a side effect. As a result, while most aerodynamic vehicles were streamlined, few streamlined vehicles were aerodynamic.

Although racecar and land speed record vehicle builders were concerned with wind resistance from before the turn of the Twentieth Century, it was not until 1934 that mass-produced streamlined passenger vehicles debuted. That year saw the introduction of the Czechoslovakian Tatra T77, as well as the American Chrysler and DeSoto Airflows.

After streamlining, the next major break in design began in 1940 when Harley Earl arranged for his styling team to visit Selfridge Air Force base. It was there that his team was given an advance look at the then-new, awe-inspiring P-38 fighter plane. The P-38 represented the ultimate in technology, speed, and beauty, and Earl commanded his men to "draw cars that look like that!" The newly inspired designers returned to their drawing boards and soon their proposals featured vehicles with

"Airex Radial" by Richard H. Arbib, 1940 (Exhibit #16)

bubble canopies, pontoon-like bodies, gaping air intakes, exposed exhaust ports, and multiple tail fins.

Although America's entry into the Second World War diverted the attention of industry and prevented the immediate incorporation of such features on production cars, new and exciting postwar technologies such as jet and rocket technology helped define the look of the postwar American automobile. Among the most prominent styling features of the 1940s and 1950s were chrome accents and dual tail fins, both of which originated at General Motors with Harley J. Earl. Soon the family car looked as capable of rocketing to the moon as it did of speeding to the grocery store.

As the modern automobile evolved, the tapered bodies and swooping fenders of the 1930s gave way to wider, slab-sided bodies with integral fenders. To give these slab-sided designs eye appeal, Earl employed chromium plated side ornamentation. Earl's use of chrome imparted a jewelry-like quality to his cars, especially when such trim was sculpted so that its surfaces would reflect into the eye of the viewer. Chrome accent trim also allowed automobiles of the mid-1950s to feature snazzy two-tone paint jobs for even more visual interest.

Dual tail fins debuted on the trailblazing 1948 Cadillac. And while they adorned Cadillacs throughout the following years, fins did not become a prominent styling feature of other American cars until 1956, the year that Vice President of Chrysler design, Virgil M. Exner, Sr., introduced the "Forward Look" for the 1957 model year. Within two years, fins rose to ludicrous heights within the industry. However, the trend for skyrocketing fins would be brief, with the only holdout by 1962 being Chrysler.

But fins were not a new concept. Already well known for its streamlining, the 1934 Tatra T77 featured a centrally placed rear "dorsal" fin which inspired the use of similar fins on several select European automobiles during the height of the Art Deco movement. Surprisingly, it was the staid Packard Automobile Company, not General Motors, who first patented both single and dual fins in America, filing their claim in 1937.

"16A" Finned Vehicle by Frank Bianchi, 1959 (Exhibit #72)

Cab-Forward Vehicle with Rear Engine
by Charles H. Stewart, 1963 (Exhibit #88)

While stylists of all generations have been concerned with form, stylists of the 1960s focused primarily on sculptured body panels to enhance their designs rather than the applied chrome trim and fancy paint schemes of previous decades. Another contrast to previous generations was a new emphasis on design continuity, which necessitated that the same design theme be employed for both the front and rear of a vehicle's design.

Yet the aerospace industry still had a great deal of influence on designers throughout the 1960s. Roofs and front ends inspired by jet airplane canopies and engine intakes remained commonplace, for example. And while the soaring tail fins of the previous decade had come and gone, new fin-like lower body panels termed "skegs" became the forefront of design. Also fashionable during the 1960s was the placement of the driver's compartment over the front wheels, a feature known as "cab-forward" design.

III: Advanced Styling

As styling progressed as a profession over the decades following 1927, specialties developed within the discipline. While most stylists were occupied with designing a vehicle's ever-evolving appearance for each new model year, a special breed of automotive designer emerged. Known as "advanced" stylists, their purpose was to predict how automobiles might look in the distant future. Advanced stylists created designs without consideration for known engineering or scientific principles hoping to set the next major trend-setting design. And as their work was meant to be inspirational, it was rarely expected to reach production. Because of the amazingly futuristic nature of these designs, they were termed "blue-sky" for having seemingly come "clear out of the blue sky." Of course the largest proponent of "blue-sky" design was Harley J. Earl.

Since their fanciful designs were never expected to reach production, advanced stylists often found themselves at great odds with corporate executives and accountants. These bean counters found it difficult to believe that paid designers would be allowed to spend

Commuter Helicopter
by Homer LaGassey, Jr., 1946 (Exhibit #33)

their time creating drawings which, in their view, were impractical and absurd. They argued that the worth of a designer was gauged by his or her ability to produce a commercially viable design. But Harley Earl felt differently, contending that if designs merely evolved slowly over time, a "breakthrough" design would likely never happen.

Throughout human history, inventors have been quick to envision creative ways to adapt new technologies to their area of specialty. Likewise, advanced stylists were inspired by technologies of other fields and used them to re-invent the automobile. At times designers even predicted cars that embodied as-yet undeveloped technologies. And still others questioned whether or not automobiles would even exist in the future.

Many of the themes found in these mid-century automobile design renderings mirrored those seen in science fiction magazines. Like advanced stylists, science fiction illustrators of the era gave little thought to contemporary engineering concerns. The strong similarity between the approach of advanced stylists and that of science fiction illustrators was likely a result of their shared desire to be ahead of their time. But unlike those of their science fiction counterparts, the visions of advanced stylists were rarely seen by the public.

Prevailing in the minds of most Americans during the mid-Twentieth Century was an almost religious conviction that, with time, there was no problem that technology could not overcome. After all, the first half of the twentieth century saw the development of many technologies that had previously been thought impossible. The Wright Brothers had fulfilled man's desire to fly. Miraculous invisible waves brought radio and television into homes. Radar could see ships and airplanes well beyond the horizon during World War II. Appliances such as vacuum cleaners, refrigerators, and clothes washers made housekeeping easier. All of these inventions pointed to the possibility that the future held wonders that could only be imagined.

In the future, cars would be comfortable, efficient, convenient, fast, and even eliminate human error.

Automated cars equipped with radar screens and proximity controls would offer a safe, hands-free traveling experience. Large, luxurious recreational vehicles would provide a means to flee civilization without leaving the comforts of home behind. Levitating cars would provide the ultimate smooth ride (and even eliminate the need to rotate tires!).

Automated Highway System
by S. Ralph Maurello, ca.1955 (Exhibit #53)

Often these creations were a result a designer's concern for anticipated future problems. One popular theory, for example, was that a growing population would result in an enormous increase in the numbers of cars on the road. This phenomenon, in turn, would eventually result in one massive traffic jam from the East Coast to the West. Designers believed that such massive gridlock would force automobiles to take to the sky. They envisioned personal helicopters and airplane-car hybrids that would free commuters from gridlock by opening the door to endless skyways. Not only would their flying cars be more convenient than conventional automobiles, they would also be quicker. The flying car concept was so popular by the late 1940s that some housing developments featured runways as well as driveways, a concept that has only recently resurfaced. Other concepts such as small commuter vehicles, single person transportation modules, and tandem-wheeled gyroscopic vehicles (half the width of conventional automobiles) were all proposed as practical ways to ease congestion without widening our nation's roads.

Over time, predictions such as automated cars, atomic-powered cars, and turbine cars were proven possible, but were eventually deemed financially or technologically impractical. Yet other seemingly far-fetched ideas, like on-board telephones, radios, televisions, and guidance systems, eventually did develop.

While our culture's unquestioning faith in technology has faded, the optimistic images of blue-sky designers continue to inspire. Some engineers still work tirelessly on designs such as flying cars, hovercraft, and gyroscopic personal transportation devices (such as the new Segway "Ginger"). Hollywood movies like The Fifth Element (1997), Minority Report (2002), and The Island (2005) continue to reinforce these popular themes. Despite the countless advances of the last century, these images serve to remind us that mankind has not advanced as far as had been predicted. And while some may regard some past designs as mere failures, there is no guarantee that these predictions of years past will not eventually come true. Only the future holds the answer!

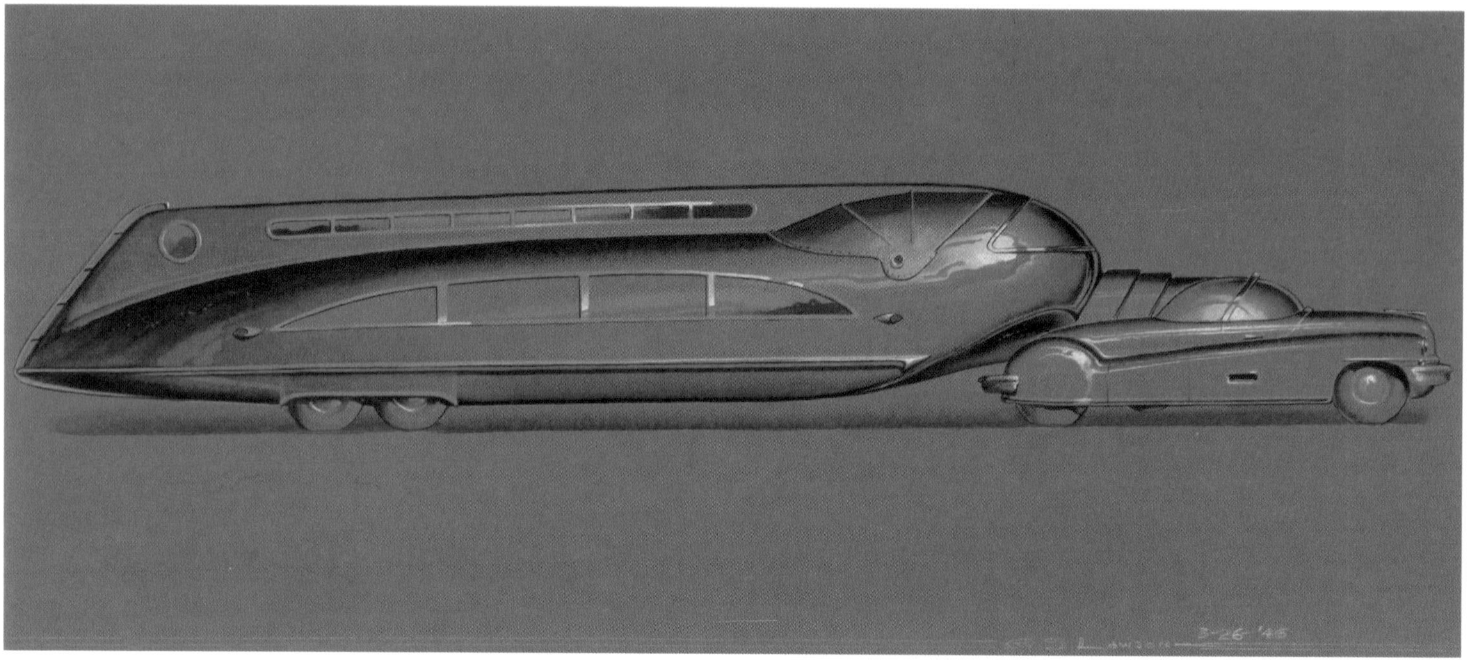

Futuristic Vehicle and Fifth-Wheel Trailer by George S. Lawson, 1946 (Exhibit #32)

Selected Exhibition Images

"Century 60" Buick Race Car by Arthur "Art" Ross, 1938 (Exhibit #9)

Bladed Front End Proposal by William Kerby, ca.1938 (Exhibit #10)

(Image courtesy of Ross family)

Dual Pontoon Buick by Arthur "Art" Ross, ca.1940 (Exhibit #17)

Patrol Car No. 6 by Jon W. Hauser I, ca.1940 (Exhibit #19)

Futuristic Red Model by Richard H. Arbib, ca.1940 (Exhibit #20)

Tucker Torpedo ¼-scale Styling Model by George S. Lawson for Preston Tucker, ca.1944-46 (Exhibit #28)

Futuristic Coupe and Women by Arthur "Art" Ross, ca.1940's (Exhibit #30)

Fisher Body Craftsman's Guild Competition Model, First Place Junior National Winner by James Mariol, 1947 (Exhibit #39)

Double Cockpit Vehicle by Ned Nickles, ca.1946 (Exhibit #36)

Micro Car Proposal by Peter W. Wozena, 1952 (Exhibit #43)

Desoto Proposal for 1959 Model Year by John R. Samsen, 1955 (Exhibit #49)

Chrysler "Imperial d'Elegance" Proposal by Chrysler Corporation Styling Staff under direction of Virgil M. Exner, Sr. (1909 - 1973), Illustrated by Thomas H. Ferriss (1926 - 2001), 1957 (Exhibit #62)

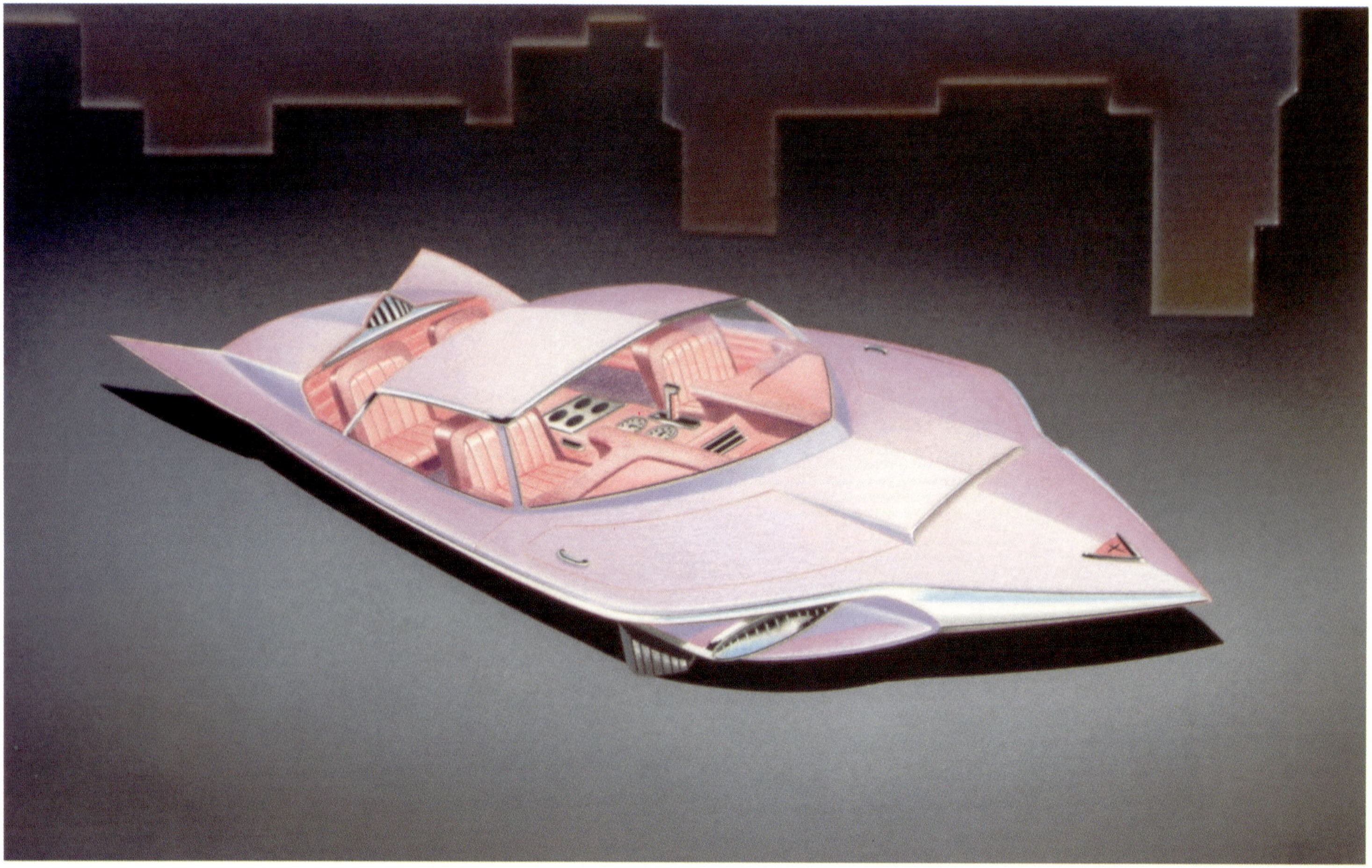

Futuristic Pink Vehicle by Peter Loda, ca.1957 (Exhibit #63)

Futuristic Automobile with Spiked Dual Fins by Louis Drake, 1958 (Exhibit #66)

Finned Automobile in India by George E. Anderson, 1959 (Exhibit #75)

Jet-inspired Cadillac Proposal by Peter W. Wozena, 1959 (Exhibit #76)

"Lizard" Utility Vehicle by Jerry L. Brochstein, ca.1958 (Exhibit #68)

Futuristic Vehicle with Finned Rear Wheel Fenders by Jerry L. Brochstein, ca.1958 (Exhibit #69)

General Motors "Shopper" proposal by Wayne K. Cherry, ca.1962 (Exhibit #82)

Cadillac V-16 Proposal by Wayne A. Kady, 1964 (Exhibit #89)

Futuristic Red Vehicle by James L. Bisignano, ca.1964 (Exhibit #90)

Family Departing Visit with Grandparents in their Commuter Helicopter by Alexis V. Lapteff, 1947 (Exhibit #38)

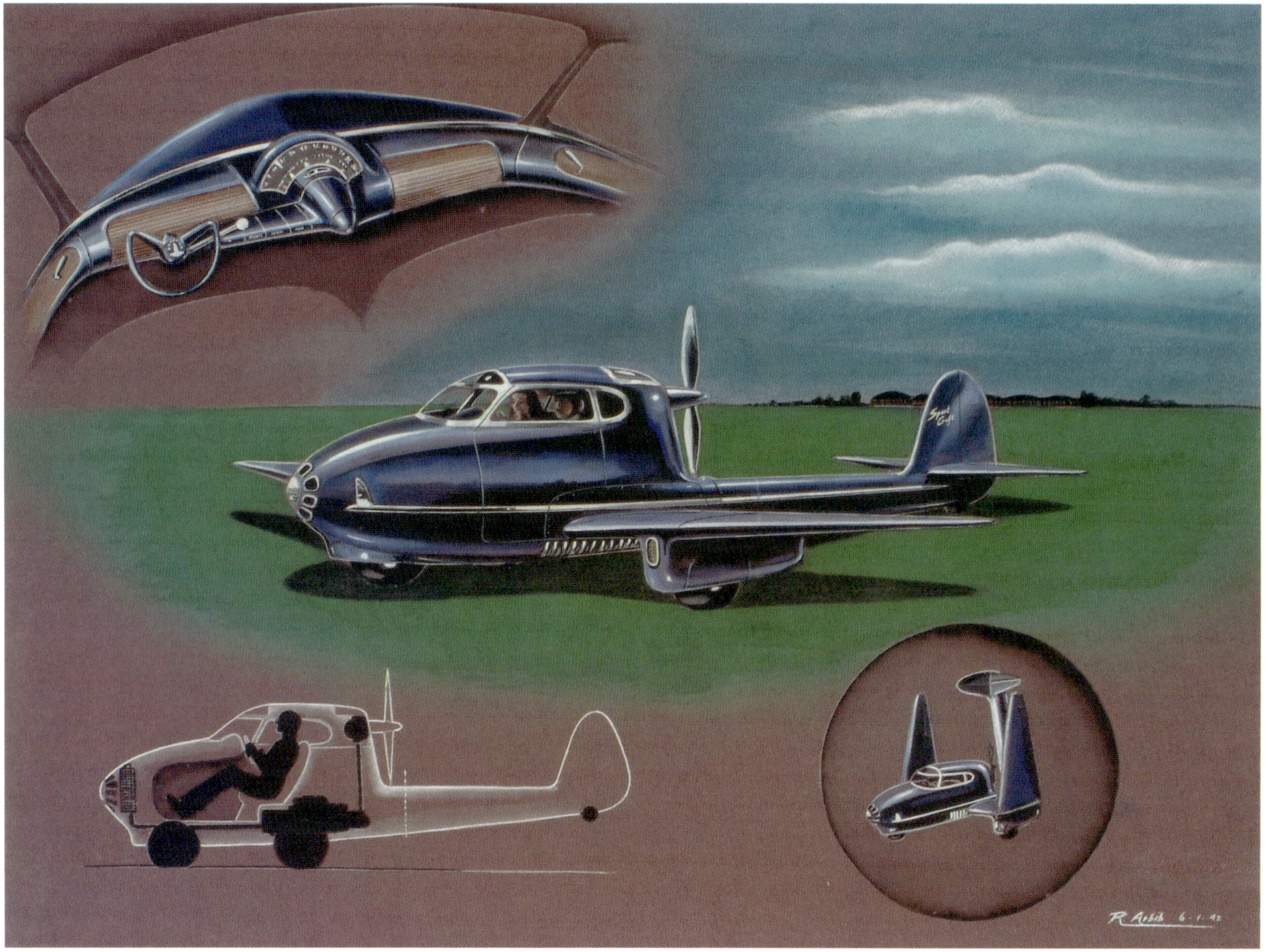

"Sport Craft" Commuter Airplane by Richard H. Arbib, 1942 (Exhibit #25)

Tandem Wheel, Gyroscopic, Atomic-Powered Flying Car by Frank R. Paul for Hugo Gernsback, 1955 (Exhibit #47)

"Radio Craft" Magazine Cover Illustration Proposal by Pearsall for Hugo Gernsback, ca.1944 (Exhibit #26)

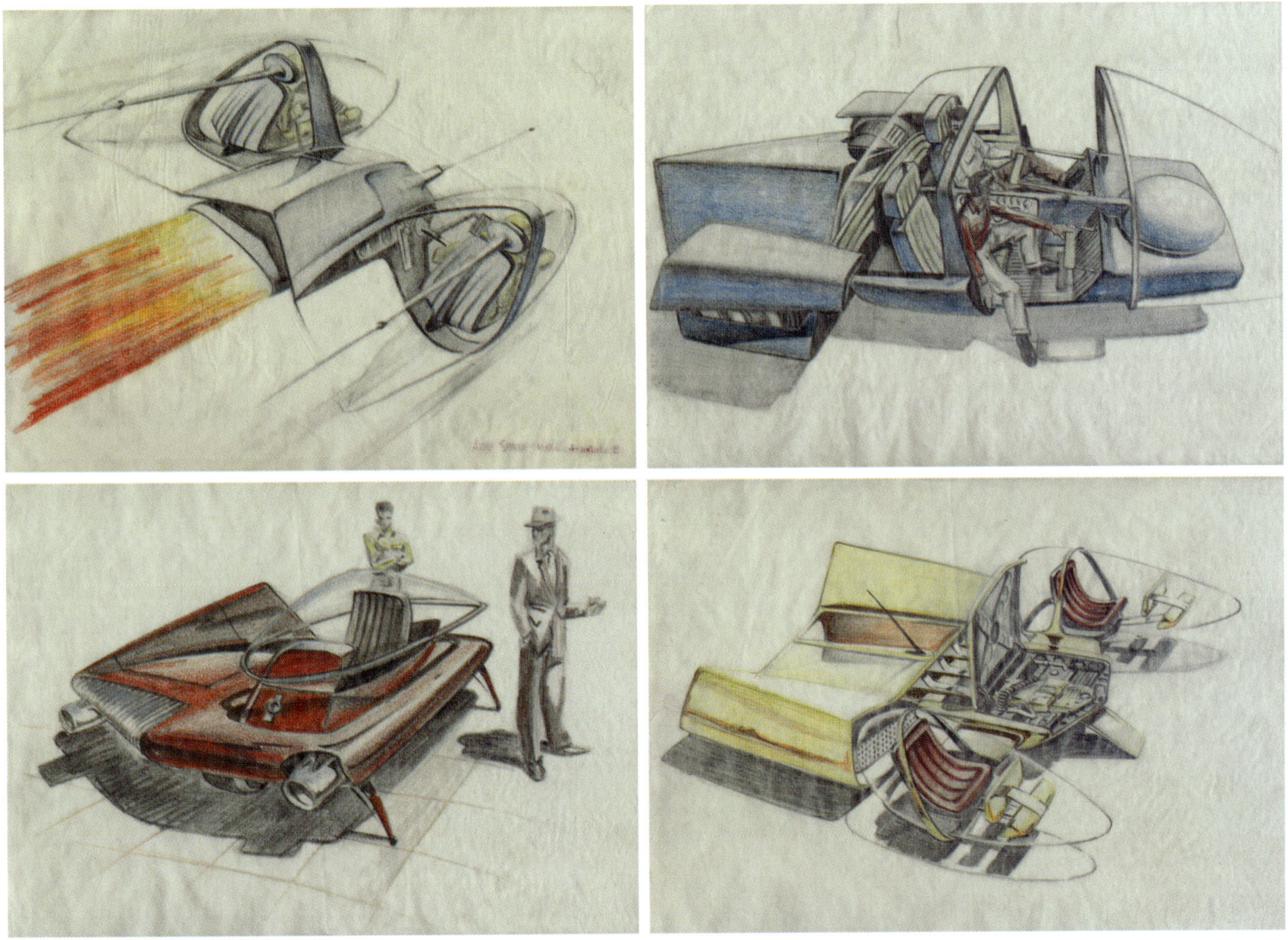

Futuristic Vehicle Concept Sketches by Sydney J. "Syd" Mead, ca.1957 (Exhibit #65)

Levitating Bus with Driver by George E. Anderson, 1959 (Exhibit #71)

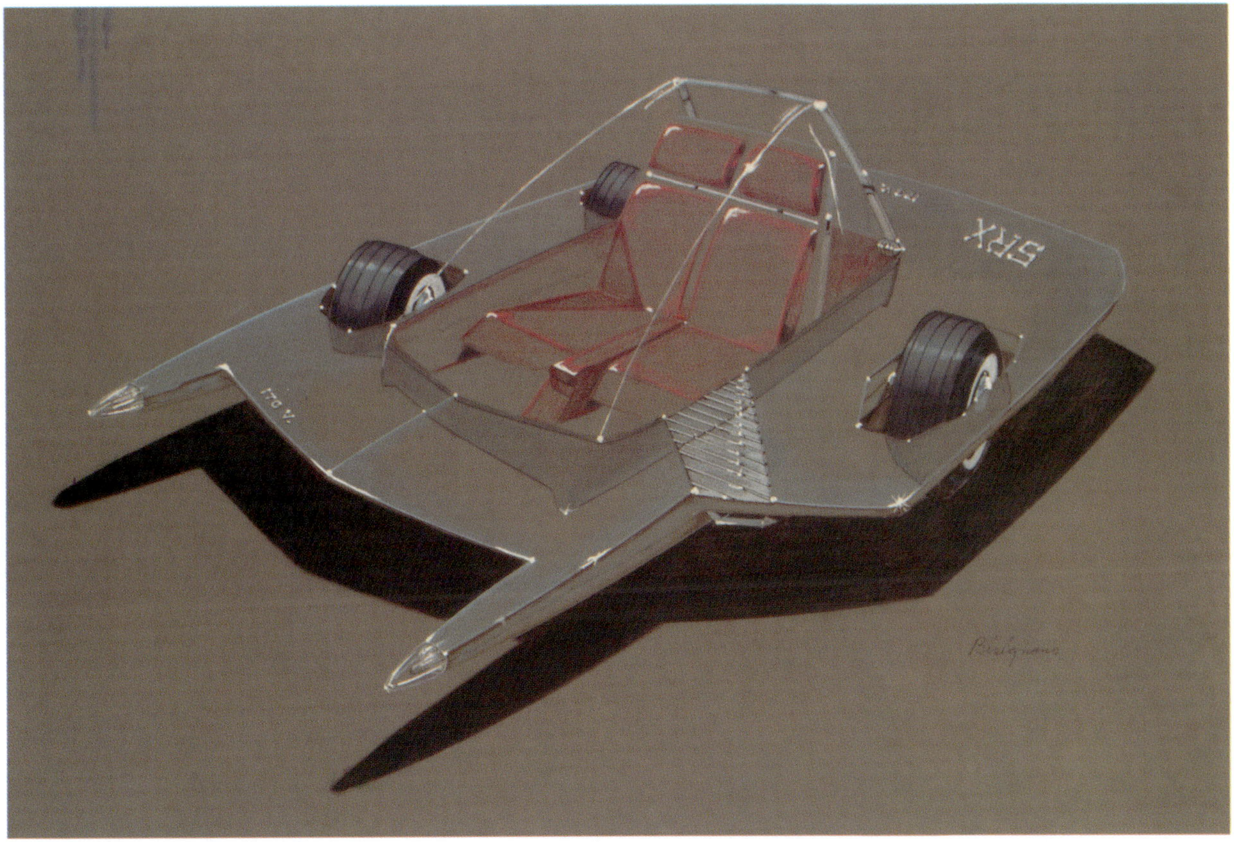

"5RX" Three-Wheel Electric Vehicle by James L. Bisignano, ca.1962 (Exhibit #86)

Ford "Gyron" Gyroscopic Concept Vehicle Proposal by Sydney J. "Syd" Mead, ca.1960 (Exhibit #78)

Jet-powered Ski Car by Robert N. Hubbach, 1961 (Exhibit #80)

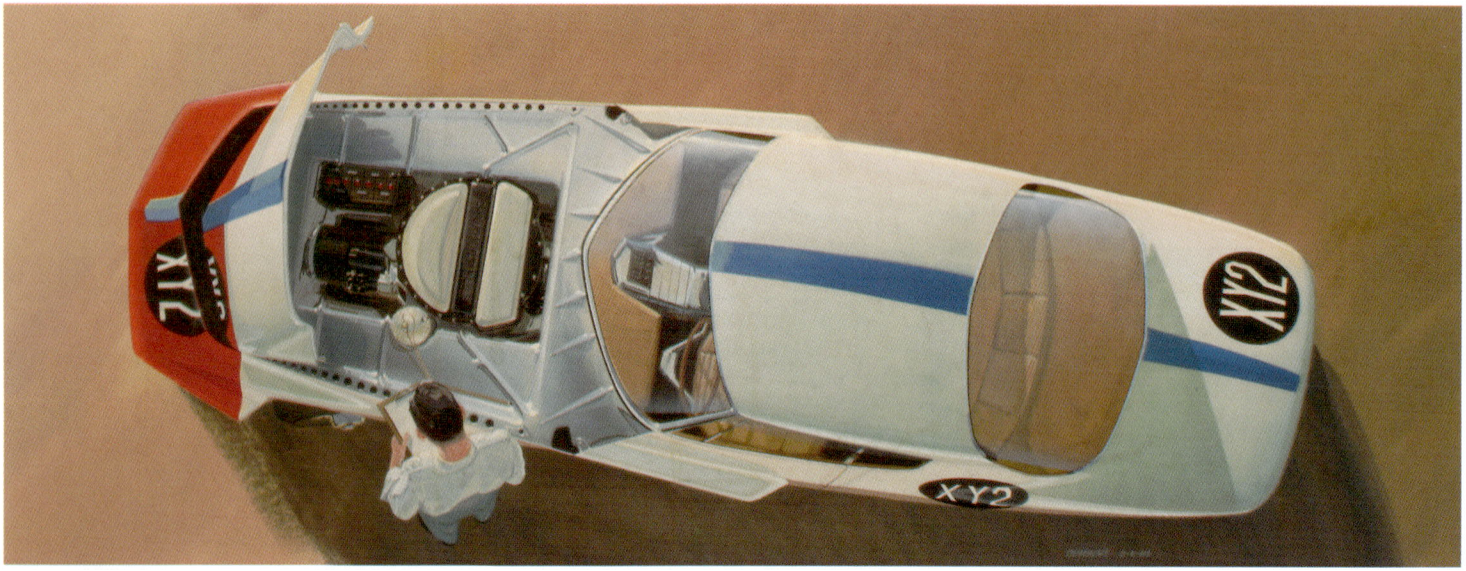

"XY2" Chrysler Turbine Car by Dave Cummins, 1960 (Exhibit #79)

(Photography by Scott Bourdon Studios, South Bend, IN)

Turbine Vehicle by Elia "Russ" Russinoff, 1962 (Exhibit #81)

"X-GM" Turbine Rail Car by Peter W. Wozena, 1964 (Exhibit #93)

Cab-Over-Engine Truck by Milton J. "Milt" Antonick, 1962 (Exhibit #84)

"Escacar" Unicycle Gyroscopic Rocket Car by Carl H. Renner, 1945 (Exhibit #29)

"Mono-Scoot" Personal Transportation Device by Richard H. Arbib, 1965 (Exhibit #95)

"Drive-Up Hotel" by Arthur C. Radebaugh, ca.1948 (Exhibit #40)

"Autoline" Road Cable System by Will Elder, ca.1955 (Exhibit #52)

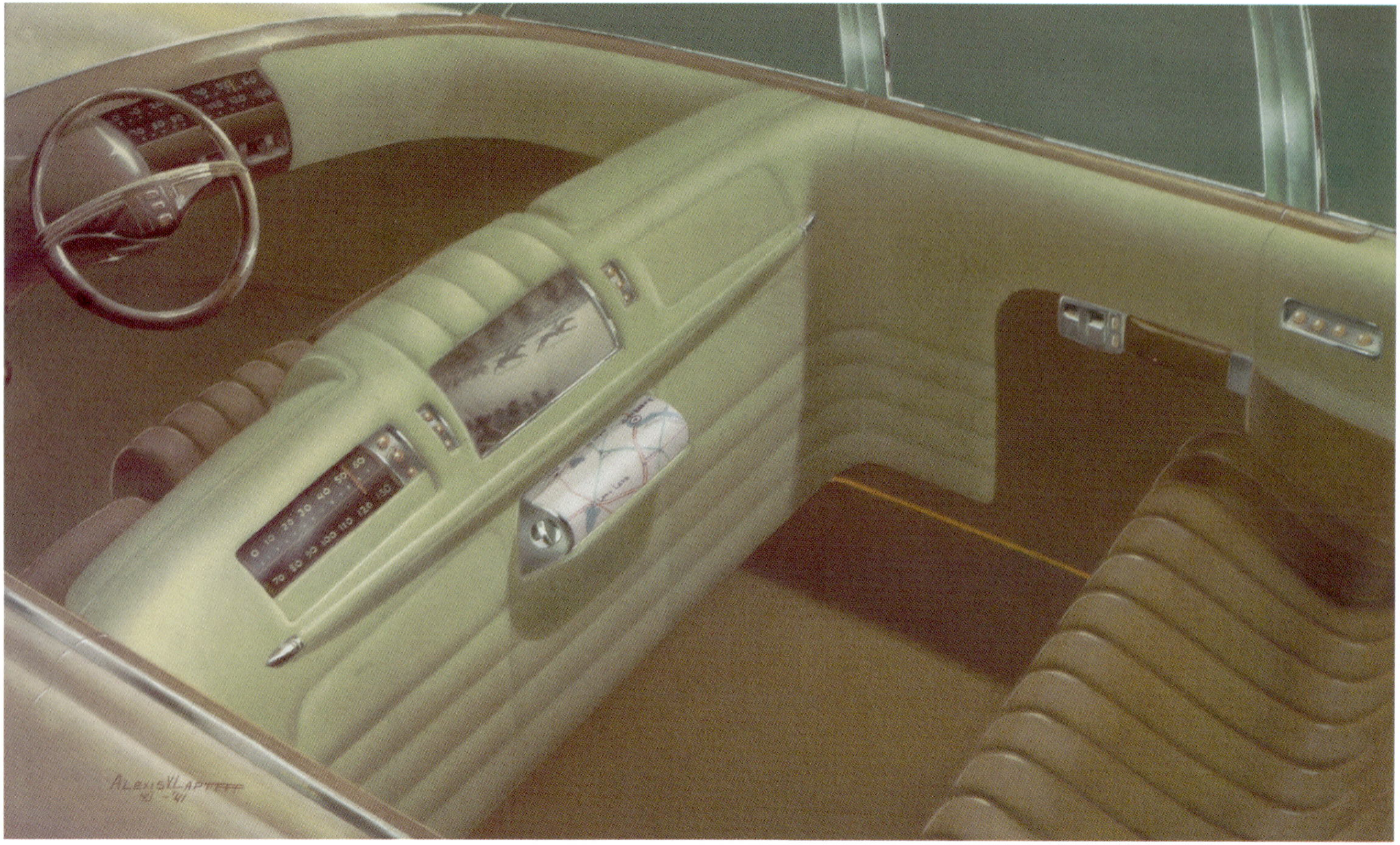

Futuristic Limousine Interior by Alexis V. Lapteff, 1941 (Exhibit #23)

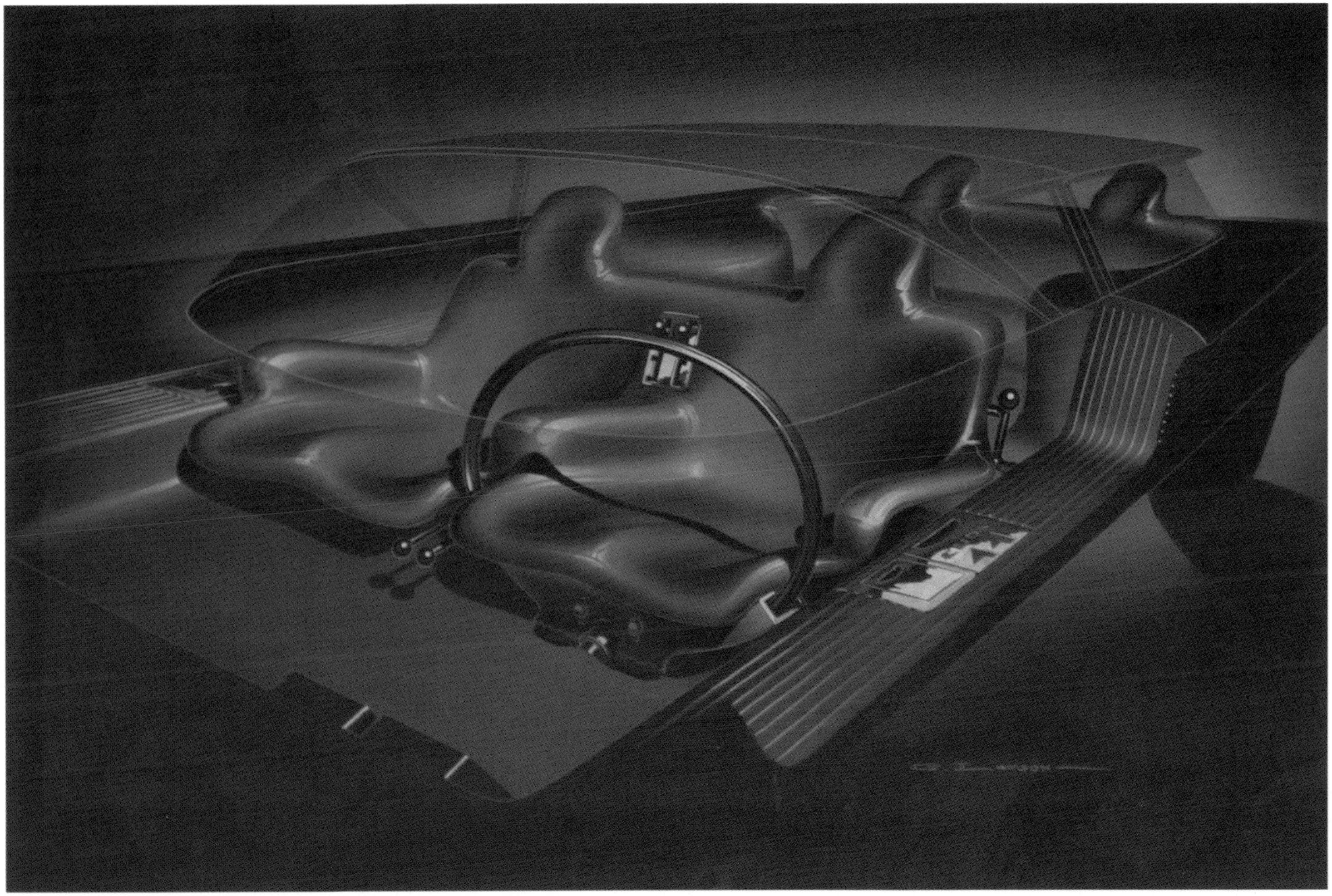

"Cuda 44" Safety Interior by George S. Lawson, ca.1958 (Exhibit #70)

Amphibious Vehicle by Richard H. Arbib, 1956 (Exhibit #56)

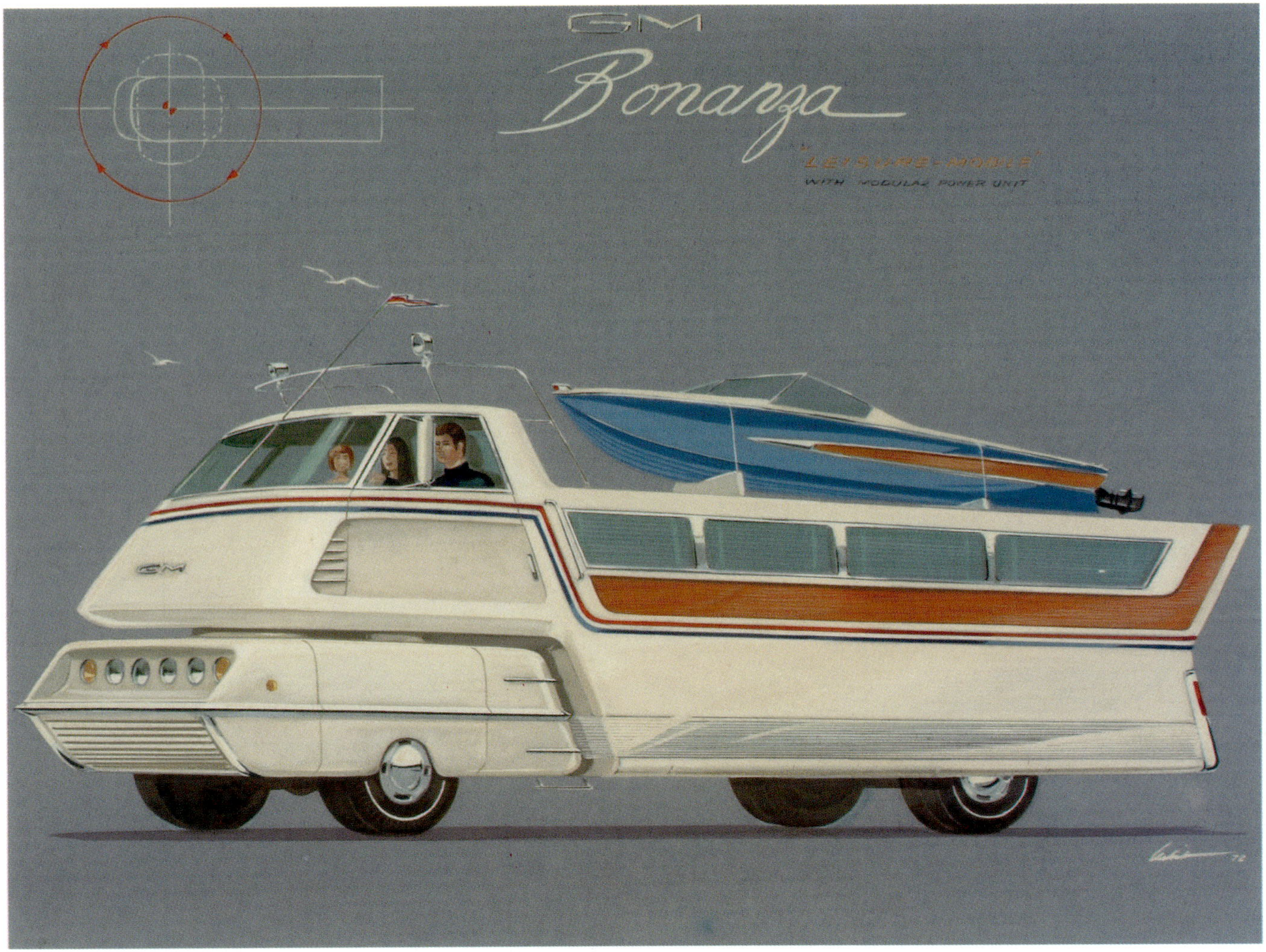

"GM Bonanza Leisure-Mobile with Modular Power Unit" by Richard H. Arbib, 1972 (Exhibit #103)

Driving Through Futures Past - Exhibit List

(Artwork from the collection of Hampton C. Wayt unless noted otherwise.)

Art and Models:

1) **Design for a Streamline Motor Car**, 1925 by Ingemar K. Rystedt (1890 - 1968), illustrated by C. C. Bratten (dates unknown), Gouache and watercolor on artist's board, 24.75"wx16"h
2) **Model Car #9**, 1933 by Norman Bel Geddes (1893 - 1958), Silver metal with plastic windows, 18"lx6"wx6"h, Collection of the Hoblitzelle Theatre Arts Library (Norman Bel Geddes Collection), Harry Ransom Humanities Research Center, The University of Texas at Austin, Mrs. Edith Letyens Bel Geddes, Executrix
3) **Tatra T77A Factory Model**, ca.1935 by Hans Ledwinka (1979 - 1967), Paul Jaray (1889 - 1974), and Erik Uberlacker (dates unknown), Painted cast aluminum, 20"lx7"wx6"h
4) **Motor Magazine Cover Art**, November 1936 by Arthur C. Radebaugh (1906 - 1974), Airbrush on artist's board, 20"wx24"h, Collection of Dave DiMartino
5) **Phantom Corsair Patent Model**, ca.1936 by Rust Heinz (1914 - 1939), modeled by Maurice Schwartz, Painted wood, 28"lx9"wx7"h
6) **Streamline Model with Fin-ette**, ca.1936 by William L. "Bill" Mitchell (1912 - 1988) for General Motors Corporation, Painted model with metal trim, 15"lx5"wx4.5"h, Collection of Henry Ford Museum, Dearborn, Michigan
7) **La Salle Proposal**, 1937 by Paul A. Meyer (1910 - 1994) for General Motors Corp., Colored pencil with gouache highlights on black paper, 17.5"wx15.5"h
8) **Safety Aircar Model**, ca.1937 attributed to Alexander R. Pribil (1880 - ca.1938), Cast aluminum, wood, metal trim, and paint, 16"lx7"wx6.5"h
9) **"Century 60" Buick Race Car**, 1938 by Arthur "Art" Ross (1913 - 1981) for General Motors Corporation, Airbrush on artist's board, 28.25"wx15.5"h
10) **Bladed Front End Proposal**, ca.1938 by William Kerby (dates unknown), Watercolor on artist's board, 17.5"wx8"h
11) **Streamlined Motorcoach #4 Model**, ca.1938 by Norman Bel Geddes (1893 - 1958), Painted metal, wood and plastic, 13"lx3"wx4"h, Collection of the Hoblitzelle Theatre Arts Library (see #2 above)
12) **Pleasure Car Model**, ca.1938 by Norman Bel Geddes (1893 - 1958), Metal, painted metal, wood, and plastic, 3"lx3"wx5"h, Collection of the Hoblitzelle Theatre Arts Library (see #2 above)
13) **Futurama Car Model (Blue)**, ca.1938 by Norman Bel Geddes (1893 - 1958), Metal, painted metal, wood, and plastic, 3"lx3"wx6.5"h, Collection of the Hoblitzelle Theatre Arts Library (see #2 above)

14) **Futurama Car Model (Green)**, ca.1938 by Norman Bel Geddes (1893 - 1958), Metal, painted metal, wood, and plastic, 3"lx3"wx9"h, Collection of the Hoblitzelle Theatre Arts Library (see #2 above)

15) **Supercharged V-Front Speedster**, 1939 by Jon W. Hauser I (1916 - 1999), Airbrush and gouache on artist's board, 15.5"wx10"h

16) **"Airex Radial"**, 1940 by Richard H. Arbib (1917 - 1995) for General Motors Corporation, Gouache and chalk on black paper, 25.4"wx31.5"h framed

17) **Dual Pontoon Buick**, ca.1940 by Arthur "Art" Ross (1913 - 1981) for General Motors Corporation, Airbrush and pastel on colored paper, 17"wx25"h, Collection of Dave DiMartino

18) **Futuristic Coupe Model with Fin-ette**, ca.1940 by William L. "Bill" Mitchell (1912 - 1988) for General Motors Corporation, Mixed media, 20"lx9"wx7.5"h, Collection of the Henry Ford Museum, Dearborn, Michigan

19) **Patrol Car No. 6**, ca.1940 by Jon W. Hauser I (1916 - 1999), Airbrush on artist's board (unsigned), 16.25"wx21"h

20) **Futuristic Red Model**, ca.1940 by Richard H. Arbib (1917 - 1995) for General Motors Corporation, Painted plaster and plexiglas, 14.5"lx6"wx3"h

21) **Finned Model**, ca.1940, Sculptor unknown, Painted hand-carved wood, 12"lx4.5"wx3.5"h

22) **Central Driver and Steering-wheel Console Proposal**, 1941 (updated 1945) by George S. Lawson (1907 - 1987), Gouache on colored paper, 17.25"wx12"h

23) **Futuristic Limousine Interior**, 1941 by Alexis V. Lapteff (1905 - 1991), Airbrush on paper, 16.5"wx10.25"h

24) **Cadillac with Integral Bumper/Grille**, ca.1941 by William L. "Bill" Mitchell (1912 - 1988) for General Motors Corporation, Watercolor and gouache on colored paper, 20"wx26"h, Collection of Henry Ford Museum, Dearborn, Michigan

25) **"Sport Craft" Commuter Airplane**, 1942 by Richard H. Arbib (1917 - 1995), Gouache on colored paper, 23"wx17"h

26) **"Radio Craft" Magazine Cover Illustration Proposal**, ca.1944 by Pearsall (dates unknown) for Hugo Gernsback (1884-1967), Gouache on artist's board, 8.75"wx12.75"h

27) **1948 Cadillac "GM Jeepster" Proposal (front and rear)**, ca.1945 by Ned Nickles (1915 - 1987) under the direction of Franklin Q. Hershey (1907 - 1997) for General Motors Corporation, Colored pencil on paper, 22.5"wx10"h

28) **Tucker Torpedo ¼-Scale Styling Model**, ca.1944-46 by George S. Lawson (1907 - 1987) for Preston Tucker, Unpainted plaster and wood, 42"lx20"wx16"h

29) **"Escacar" Unicycle Gyroscopic Rocket Car**, 1945 by Carl H. Renner (unsigned) (1923 - 2001) for General Motors Corporation, Gouache on artist's board, 20.5"wx14"h

30) **Futuristic Coupe and Women**, ca.1940's by Arthur "Art" Ross (1913 - 1981) for General Motors Corporation, Airbrush on colored paper, 24"wx13"h

31) **Finned Single Cockpit Vehicle Heading to Town**, 1946 by Carl H. Renner (1923 - 2001) for General Motors Corporation, Gouache and airbrush on colored paper, 23.5"wx16"h

32) **Futuristic Vehicle and Fifth-Wheel Trailer**, 1946 by George S. Lawson (1907 - 1987), Gouache on colored paper, 24.75"wx11.5"h

33) **Commuter Helicopters (two pieces)**, 1946 by Homer LaGassey, Jr. (b. 1924), Colored pencil on tissue (date in pen), each 15.75"wx13.5"h

34) **Sleek Sedan Driving to Winkler Mill Farm**, ca.1946 by Ned Nickles (1915 - 1987) for General Motors Corporation, Watercolor on artist's board, 20"wx16"h, Collection of the Petersen Automotive Museum, Gift of Franklin Q. Hershey

35) **"Le Prix" Fastback Finned Sedan**, ca.1940's by Jon W. Hauser I (1916 - 1999), Mixed media on colored paper, 20.25"wx13.5"h

36) **Double Cockpit Vehicle**, ca.1946 by Ned Nickles (1915 - 1987) for General Motors Corporation, Gouache and airbrush on artist's board, 20"wx13.25"h

37) **"Aerolite" Dual Cockpit Vehicle**, ca.1946 by Carl H. Renner for General Motors Corporation, Watercolor, pencil and gouache, 25.5"wx19.5"h, Collection of Joe Molina

38) **Family Departing Visit with Grandparents in their Commuter Helicopter**, 1947 by Alexis V. Lapteff (1905 - 1991), Gouache on artist's board, 16"wx10.25"h

39) **Fisher Body Craftsman's Guild Competition Model, First Place Junior National Winner**, 1947 by James Mariol (b. 1930), Painted hand-carved wood with aluminum trim, approx. 14"lx6"wx6"h

40) **"Drive-Up Hotel"**, ca.1948 by Arthur C. Radebaugh (1906 - 1974), Airbrush on artist's board, 19.5"wx12"h

41) **General Motors LeSabre Design Proposal**, 1949 by Homer LaGassey, Jr. (b.1924) for General Motors Corporation, Colored pencil and gouache on colored paper, 19"wx12"h

42) **Interior Proposal for the Ford X-100 Show Car**, 1952 by John Najjar (b.1918) for Ford Motor Company, Colored pencil on tissue, 25"wx19"h

43) **Micro Car Proposal**, 1952 by Peter W. Wozena (b.1914) for General Motors Corporation, Colored pencil and gouache on tissue, 24.5"wx18"h

44) **Truck featuring Automobile Design Elements**, 1952 by Charles M. "Chuck" Jordan (b.1927) for General Motors Corporation, Gouache on artist's board, 20"wx30"h, Collection of Charles M. "Chuck" Jordan

45) **Firebird I Turbine Car Executive Model**, 1954 by the General Motors Styling Staff under direction of Harley J. Earl (1893 - 1969), Painted plaster, wood, and metal, 40"lx14.5"wx10"h

46) **Opel Race Cars**, ca.1954 by Ronald C. Hill (b.1934), Watercolor on watercolor paper (unsigned), 7"wx6"h

47) **Tandem Wheel, Gyroscopic, Atomic-Powered Flying Car**, 1955 by Frank R. Paul (1884 - 1963) for Hugo Gernsback (1884-1967), Gouache and ink on board, 16.75"wx11"h

48) **Tandem Wheel Gyroscopic Vehicles**, 1955 by Frank R. Paul (1884 - 1963) for Hugo Gernsback (1884 - 1967), Mixed Media on board, 24.5"wx17"h

49) **Desoto Proposal for 1959 Model Year**, 1955 by John R. Samsen (b.1927) for Chrysler Corporation, Airbrush and gouache on paper mounted on board, 15.5"wx9.75"h

50) **Ford "Nucleon" Atomic-Powered Car Model**, 1955 by James R. Powers (b.1934) and Ford Design Staff, Painted fiberglass, 75"lx29"wx16"h, Collection of the Henry Ford Museum, Dearborn, Michigan

51) **1958 Ford "Volante" Air-Ducted Levitating Car Model**, 1955 by James R. Powers (b.1934) and Ford Design Staff, Painted fiberglass, 80"lx32"wx16"h, Collection of the Henry Ford Museum, Dearborn, Michigan

52) **"Autoline" Road Cable System**, ca.1955 by Will Elder (b.1921), Gouache and ink on board (unsigned), 16.5"wx9.75"h

53) **Automated Highway System**, ca.1955 by S. Ralph Maurello (1911 - 1981), Airbrush on artist's board, 13"wx18.5"h

54) **Dodge "Royal" Proposal**, 1956 by Don Reid under Direction of Virgil M. Exner, Sr. (1909- 1973) for Chrysler Corporation, Airbrush and gouache on colored paper, 104"wx56"h, Collection of Charles S. Hayes, South Bend, Indiana

55) **Firebird II Turbine Car Model**, 1956 by the General Motors Styling Staff under the direction of Harley J. Earl (1893 - 1969), Painted plaster, wood, plexiglas and metal, 15"lx4.5"wx3.5"h

56) **Amphibious Vehicle**, 1956 by Richard H. Arbib (1917 - 1995), Gouache on colored paper, 15.5"wx12.5"h

57) **Dual Cockpit Buick**, 1956 by Peter W. Wozena (b.1914) for General Motors Corporation, Colored pencil with gouache highlights on tissue, 24"wx16"h

58) **Rear 3/4 Proposal with Fin**, 1956 by Peter W. Wozena (b.1914) for General Motors Corporation, Colored pencil with gouache highlights on tissue

59) **Levitating Automobiles**, 1957 by Jerry L. Brochstein (b.1937), Colored pencil with gouache highlights on tissue, 22"wx15"h

60) **Front End Proposals (two pieces)**, 1957 by Peter W. Wozena (b.1914) for General Motors Corporation, Colored pencil with gouache highlights on tissue, each 13.75"wx16.75"h

61) **Rear Wing Proposal**, 1957 by Milton J. "Milt" Antonick (b.1936), Colored pencil with gouache highlights on colored paper, 24.5"wx19"h

62) **Chrysler "Imperial d'Elegance" Proposal**, 1957 by Chrysler Corporation Styling Staff under direction of Virgil M. Exner, Sr. (1909 - 1973), Illustrated by Thomas H. Ferriss (1926 - 2001), Airbrush on artist's board, 26.75"wx15.75"h

63) **Futuristic Pink Vehicle**, ca.1957 by Peter Loda (b.1936), Mixed media on artist's board (unsigned), 14.25"wx8.75"h

64) **Finned Blue Model**, ca.1957 by Milton J. "Milt" Antonick (b.1936), Painted hand-carved wood, aluminum, and plastic, 18.5”lx7”wx4.5”h

65) **Futuristic Vehicle Concept Sketches (four pieces)**, ca.1957 by Sydney J. "Syd" Mead (b.1933), Colored pencil on paper, each 11”wx8.25”h

66) **Futuristic Automobile with Spiked Dual Fins**, 1958 by Louis Drake (dates unknown), Colored pencil on paper, 22”wx17.5”h

67) **Piasecki "Airgeep"**, 1958 by the Piasecki Helicopter Corporation, Illustrated by Fred Freeman (1906 - 1988), Casein and wash on artist's board, 26.5”wx9”h

68) **“Lizard” Utility Vehicle**, ca.1958 by Jerry L. Brochstein (b.1937), Gouache and colored pencil on colored paper, 23”wx16.5”h

69) **Futuristic Vehicle with Finned Rear Wheel Fenders**, ca.1958 by Jerry L. Brochstein (b.1937), Gouache and colored pencil on colored paper, 23”wx17.5”h

70) **“Cuda 44” Safety Interior**, ca.1958 by George S. Lawson (1907 - 1987) for American Motors Corporation, Airbrush, gouache, and colored pencil on black paper, 28”wx15.75”h

71) **Levitating Bus with Driver**, 1959 by George E. Anderson (b.1938), Colored pencil with gouache highlights, 17.5”wx17.5”h

72) **“16A” Finned Vehicle**, 1959 by Frank Bianchi (1918 - 2005) for Chrysler Corporation, Acrylic on artist's board, 24.5”wx14.5”h

73) **“Experior” Superior Experimental Six-Wheel Vehicle**, 1959 by Galen Wickersham (b.1933), Colored pencil with gouache highlights on colored paper

74) **“Baroque Beta Mach 7” Anti-Gravity Limousine**, 1959 by Sydney J. "Syd" Mead (b.1933) for Ford Motor Company, Gouache with decal lettering on artist's board, 46”wx33”h

75) **Finned Automobile in India**, 1959 by George E. Anderson (b.1938), Watercolor on artist's board, 29.25”wx19.25”h

76) **Jet-inspired Cadillac Proposal**, 1959 by Peter W. Wozena (b.1914) for General Motors Corporation, Colored pencil with gouache highlights on tissue, 25”wx15.5”h

77) **Ford “Gyron” Gyroscopic Concept Vehicle Speed Model** (on non original stand), ca.1960 by the Ford Styling Staff, Painted fiberglass, 20”lx11.5”wx3.5”h

78) **Ford “Gyron” Gyroscopic Concept Vehicle**, 1960 by Sydney J. "Syd" Mead (b.1933) for Ford Motor Company, Gouache and airbrush on artist's board, 38.5”wx26”h

79) **“XY2” Chrysler Turbine Car**, 1960 by Dave Cummins (b.1934) for Chrysler Corporation, Gouache on artist's board, 46.5”wx16.25”h

80) **Jet-powered Ski Car**, 1961 by Robert N. Hubbach (b.1938), Gouache on artist's board, 23”wx17”h

81) **Turbine Vehicle**, 1962 by Elia "Russ" Russinoff (b.1930) for General Motors Corporation, Airbrush and gouache on colored paper, 84”wx42”h

82) **General Motors “Shopper” Proposal**, 1962 by Wayne K. Cherry (b.1937) for General Motors Corporation, Mixed media on artist's board, 23”wx13”h

83) **Angular Double Cockpit Automobile**, 1962 by John Perkins (b.1940) for General Motors Corporation, Flowmaster ink with gouache highlights on tissue, 15.25”wx12.75”h

84) **Cab-Over-Engine Truck**, 1962 by Milton J. "Milt" Antonick (b.1936) for Studebaker Corporation, Mixed media on colored paper, 24.5”wx17.5”h

85) **Firebird IV Turbine Car**, 1962 by Charles H. Stewart (b.1935) for General Motors Corporation, Flowmaster ink on tissue, 26.25”wx13”h

86) **“5RX” Three-Wheel Electric Vehicle**, ca.1962 by James L. Bisignano (b.1939), Gouache on colored paper, 20”wx12.5”h

87) **Plymouth Barracuda Proposal**, 1963 by John R. Samson (b.1927) for Chrysler Corporation, Mixed media on paper, 22.25”wx16.5”h

88) **Cab-Forward Vehicle with Rear Engine**, 1963 by Charles H. Stewart (b.1935) for General Motors Corporation, Flowmaster ink on tissue, 16.75”wx10.25”h

89) **Cadillac V-16 Proposal**, 1964 by Wayne A. Kady (b.1939) for General Motors Corporation, Flowmaster ink on tissue, 35”wx24”h, Collection of Gary MacNorius

90) **Futuristic Red Vehicle**, 1964 by James L. Bisignano (b.1939) for General Motors Corporation, Flowmaster ink with gouache highlights on tissue, 24.75”wx18.75”h

91) **“A-11” Jet-Powered Car**, 1964 by Robert Ackerman (b. 1935) for General Motors Corporation, Ink on tissue, 16”wx7”h

92) **“B-70” Automobile**, 1964 by Charles H. Stewart (b.1935) for General Motors Corporation, Flowmaster ink on tissue, 26”wx16”h

93) **“X-GM” Turbine Rail Car**, 1964 by Peter W. Wozena (b.1914) for General Motors Corporation, Flowmaster ink and gouache on tissue, 22”wx15.5”h

94) **Hovercraft**, ca.1964 by Ben Delphia (b.1938), Pen on paper, 6.25”wx6.25”h

95) **"Mono-Scoot" Personal Transportation Device**, 1965 by Richard H. Arbib (1917 - 1995), Gouache on colored paper, 22.75”wx17”h

96) **Magnetic Levitating Bus (two panels)**, 1965 by Frederick J. "Bud" Magaldi (b.1944), Gouache on artist's board, each 23.5"wx19"h

97) **Fisher Body Craftsman's Guild Competition Model, State Third Place Award**, 1966 by Terry P. Graboski (b.1947), Painted carved wood with rubber wheels, approx. 14"lx6"wx6"h, Collection of Terry P. Graboski

98) **Magnetic Levitating (MagnaLev) Automated Car**, 1967 by Al Gerstenberger (b. 1942), Mixed media on paper, 23.5"wx17.5"h

99) **Fisher Body Craftsman's Guild Competition Model, South East Regional Division First Place Award**, 1967 by Terry P. Graboski (b.1947), Painted carved wood with rubber wheels, approx. 14"lx6"wx6"h, Collection of Terry P. Graboski

100) **Fisher Body Craftsman's Guild Competition Model, National Fifth Place Award**, 1968 by Terry P. Graboski (b.1947), Painted carved wood with rubber wheels, approx. 14"lx6"wx6"h, Collection of Terry P. Graboski

101) **"Cyclar" Enclosed Two-Passenger Motorcycle/Car**, 1969 by Richard H. Arbib (1917 - 1995), Gouache on colored paper, 23.5"wx17"h

102) **Motor Home Proposal**, 1969 by David D. Rieden (b. 1941) for General Motors Corporation, Gouache on artist's board, 29"wx19"h

103) **"GM Bonanza Leisure-Mobile with Modular Power Unit" Recreational Vehicle with Motorboat**, 1972 by Richard H. Arbib (1917 - 1995), Gouache on colored paper, 23.5"wx17.5"h

Automobiles:

104) **1933 Chrysler Trifon**, Collection of DaimlerChrysler Corporation

105) **1936 Stout Scarab**, Collection of Larry Smith

106) **1948 Davis**, Collection of the Petersen Automotive Museum Foundation

107) **1948 Tucker (Serial #30)**, Preston Tucker's Personal Car, Collection of the Petersen Automotive Museum

108) **1954 Pontiac Bonneville Special**, Collection of Roger Willbanks

109) **1956 Nash Metropolitan Astra-Gnome**, Collection of James Valentine, Metropolitan Pit Stop